American FARM TRACTOR TRADEMARKS

Encyclopedia of Tractor Trademarks 1870s–1960s

J. I. Case 1930

C. H. Wendel

First published in 1994 by Motorbooks International Publishers & Wholesalers, PO Box 2, 729 Prospect Avenue, Osceola, WI 54020 USA

© Charles H. Wendel, 1994

All rights reserved. With the exception of quoting brief passages for the purposes of review no part of this publication may be reproduced without prior written permission from the Publisher

Motorbooks International is a certified trademark, registered with the United States Patent Office

The information in this book is true and complete to the best of our knowledge. All recommendations are made without any guarantee on the part of the author or Publisher, who also disclaim any liability incurred in connection with the use of this data or specific details

We recognize that some words, model names and designations, for example, mentioned herein are the property of the trademark holder. We use them for identification purposes only. This is not an official publication

Motorbooks International books are also available at discounts in bulk quantity for industrial or sales-promotional use. For details write to Special Sales Manager at the Publisher's address

Library of Congress Cataloging-in-Publication Data

Wendel, C. H. (Charles H.)
   American farm tractor trademarks, 1840-1993/C. H. Wendel.
      p. cm. —(Motorbooks International Crestline)
   Includes index.
   ISBN 0-87938-931-1
   1. Farm tractors—United States—Trademarks.
   2. Farm tractors—United States—History. I. Title.
II. Series: Crestline series.
TL233. W457   1994
629.225—dc20   94-32795

On the front cover: The restored experimental John Deere Model C, serial number 200109, owned by the Keller family of Forest Junction, Wisconsin. *Andrew Morland*

On the back cover: Trademarks from J. I. Case 1930, International Harvester 1944, Ford-Ferguson 1943, and Caterpillar 1916.

Printed and bound in the United States of America

Contents

Introduction 4
Alphabetical Listing of Tractor Trademarks 6
Index 127

Introduction

Although we made no attempt to keep track of the hours involved researching this book, it involved page-by-page research of the weekly *Patent Office Gazette* from 1872 to 1960. This totals to well over 4,500 issues! Conservatively, we estimate some 2,000 hours of research for this project. However, it has been a most interesting excursion, and has brought forward numerous companies for which no other history has been found, despite a quarter century of research work into the history of the farm tractor. In this light then, it is a great pleasure for us to present the completed project. Hopefully, there will be folks who will recognize certain companies and be able to shed more light on their activities.

One of the great problems for this book has been to obtain good reproductions of the material printed in the *Patent Office Gazette*. Some appear to have been printed on contract, while others may have been produced by the Government Printing Office. Regardless of the source, many of the trademarks are not of the quality one would desire, particularly since the author is a letterpress printer. Yet, there has been no other choice but to enhance each print as much as possible through all sorts of darkroom chicanery. Even with many years of darkroom experience, this has not always resulted in the quality we would have liked.

In going through this book, it is to be remembered that the majority of the marks shown are those published for opposition. It seemed outside the scope of our research to follow these marks to see when or if they were actually granted, but the vast number of them were used by the companies who filed them.

ГУСЕНИЦА
Caterpillar's Russian trademark 1927

Some of the marks illustrated in this book are still active, and these, as well as all the marks shown, are presented solely in a historical sense. Their presence does not constitute any sales pitch for, or any diminution of, the companies represented by these trademarks.

Massey-Harris 1951

The dates below each trademark are the dates the trademark application was filed. However, in many cases firms were using the trademarks years before the filing date.

If this book achieves any status whatever in its scope or its historical sense, then the Author will feel well rewarded for the effort. Comments regarding this book will be appreciated, and possibly, we will compile other titles of this series to include such subjects as automobiles, farm and industrial equipment, and internal combustion engines.

Please direct your comments to the author, in care of Motorbooks International.

C. H. Wendel

C. H. Wendel

Alphabetical Listing of Tractor Trademarks

Advance-Rumely Company
LaPorte, Indiana
Advance-Rumely was the reorganized version of the M. Rumely Company, established in 1853. The famous Rumely OilPull tractors continued as a hallmark until the company was acquired by Allis-Chalmers in 1931.

1918

1928

1930

**Advance Thresher Company
Battle Creek, Michigan**

Advance Thresher Company was an old, established company engaged in building grain threshers, followed by steam traction engines, and even a few tractors. This company was bought out by M. Rumely Company, LaPorte, Indiana, in 1911.

1906

**Albaugh-Dover Company
Chicago, Illinois**

Albaugh-Dover built the Square Turn tractor at Lincoln, Nebraska. The Square Turn was of an unusual design and manifested an uncanny resemblance to the as-yet undeveloped row-crop tractor. The firm began its existence as the Kenney-Colwell Company at Norfolk, Nebraska. Albaugh-Dover bought the firm in 1916, and it was sold at a sheriff's sale in 1925.

1909

SquareTurn

1917

S. L. Allen & Company, Inc.
Philadelphia, Pennsylvania
The Planet Jr. implements were widely known for many years, particularly for truck farming and gardening. The trademark shown here was for its small garden tractors, first used in 1920.

Planet Jr.

1920

Allis-Chalmers Manufacturing Company
Milwaukee, Wisconsin
Allis-Chalmers resulted as a merger of several large manufacturing concerns in 1903. Although a major industrial manufacturer, the company entered the tractor business in 1914. Over the years, the product line expanded to include a full line of farm equipment. Allis-Chalmers was merged with Deutz in 1985 to form the Deutz-Allis tractor line.

ALLIS - CHALMERS

1947

ALLIS-CHALMERS

1949

1950

ALLIS-CHALMERS
1955

1956

**American Engine & Tractor Company
Charles City, Iowa**
Beginning in 1918, American Engine & Tractor offered the American 15-30 model. Production ended in 1920, with the company possibly being reorganized as American Tractor & Foundry Company.

1918

American Tractor & Foundry Company
Charles City, Iowa

American Tractor & Foundry emerged in 1920 as builders of the Americo tractor. Few details of this firm can be found. From all indications, this company lasted for only a short time.

American Tractor Corporation
Churubusco, Indiana

The Terratrac crawler tractors quickly developed into a very popular line with numerous innovative features. The Terratrac emerged in 1949, and the company was merged with J. I. Case Company in 1956.

1920

TERRATRAC

1949

1953

American Tractor Corporation
Peoria, Illinois
Little is known of this company except that it began building tractors in 1918. Since it is absent from the industry listings, the assumption is that production was indeed, limited.

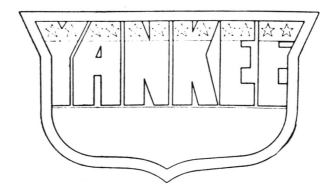

1920

Aro Tractor Company
Minneapolis, Minnesota
Aro began building tractors in 1921, but few traces of this firm can be found in the industry listings.

1921

Aultman-Taylor Company
Mansfield, Ohio
This company was organized in 1890 to build engines and threshers. It began building large tractors about 1910, and was bought out by the Advance-Rumely Company of LaPorte, Indiana, in 1924.

1914

Auto Power Company
Chicago, Illinois

"Helping Henry" was actually a device that permitted belt power to be delivered from the ever-present Ford Model T automobile. A great many of these devices were sold for a time as a substitute for tractor power.

"HELPING HENRY"

1916

Autopower Company
Detroit, Michigan

Numerous companies offered power takeoff devices that permitted an automobile to drive various farm machines. Others converted the family car into a tractor of sorts; the Autopow was one such device.

AUTOPOW

1916

Avery Company
Peoria, Illinois

The Avery Company registered its well-known "Bulldog" in 1907, along with the Avery trademark. The Avery Track-Runner was a small crawler tractor of the early 1920s. Avery went bankrupt in 1924, but was subsequently reorganized.

1907

1907

1921

**B. F. Avery & Sons Company
Louisville, Kentucky**

B. F. Avery built farm equipment at Louisville from 1845 onward. The company made little effort to enter the tractor business until the 1930s, but its later designs were unique and innovative. The company was absorbed by Minneapolis-Moline in 1951.

1941

WITTE

1950

TRU-DRAFT

1941

1941

Badley Tractor Company
Portland, Oregon
Little is known of the Angleworm tractor that first appeared about 1934 and remained viable for a few years. The Angleworm was a unique crawler design only about 3ft in height.

1937

Baines Engineering Company
Canal Dover, Ohio
While the trademark application for the Cultractor indicates use since March 1920, virtually no trace of this unit has been located.

1931

William Stimson Barnes
Chicago, Illinois
The first use claimed for the Nu-Horse trademark is in 1915, thus indicating what was likely the early beginnings of this endeavor. However, our research has found no listings for this tractor.

1915

NUHORSE

1915

Bean Spray Pump Company
San Jose, California
As the company name indicates, this firm was a manufacturer of spraying apparatus, and eventually got into the tractor business in a small way. The tractor venture was not eminently successful.

1917

Belle City Manufacturing Company
Racine, Wisconsin
Belle City began building various farm machinery in the 1890s, with threshing machines, silo fillers, and feed cutters being its principal business. Eventually the company built crawler attachments, particularly for the Fordson tractors.

1925

1918

Beltrail Tractor Company
St. Paul, Minnesota

Between 1918 and 1920, Beltrail built crawler tractors. The company was organized, reorganized, and out of business in only a couple of years.

1912

C. L. Best Tractor Company
San Leandro, California

This company was organized in 1910 to build crawler tractors. Within a short time it rose to dominance in its field. In 1925, Best merged with the Holt Manufacturing Company to form Caterpillar Tractor Company.

1913

1919

BEST

TRACKLAYER

1921

1921

BEST TRACKLAYER TRACTORS

1924

Blumberg Motor Manufacturing Company
San Antonio, Texas
Blumberg offered 9-18 and 12-24 tractor models between 1915 and 1924. Even though it was a relatively small company, these tractors featured engines built within the Blumberg factory.

BLUMBERG STEADY PULL TRACTOR

1921

David Brown Corporation, Ltd.
London, England
Although of British origin, the David Brown tractors have received wide acceptance in the United States, and of course, throughout the world.

DAVID BROWN

1958

Buckeye Manufacturing Company
Anderson, Indiana
This company started building tractors in 1912, but the first "Trundaar" model appeared in late 1917. Production ended in the early 1920s.

Trundaar

1918

Buffalo Pitts Company
Buffalo, New York
Although the two trademarks shown date back to 1876 and 1893, the brothers Pitts had begun business in the 1830s with America's first successful grain thresher. Buffalo Pitts began building tractors about 1910, and continued to about 1920.

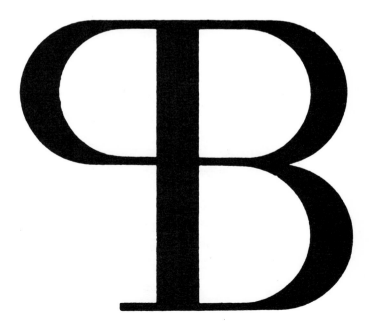

1899

BUFFALO PITTS

1912

Bull Tractor Company
Minneapolis, Minnesota
When the Bull tractor appeared in 1913, it revolutionized the farm tractor industry, since it was the first commercially successful small tractor. Although it was a short-lived venture, the Bull was certainly a forerunner of things to come.

BULL

1913

Bullock Tractor Company
Chicago, Illinois
Bullock Tractor Company was organized in 1913 from the Western Implement & Motor Company. The latter had begun using the Creeping Grip name already in 1911. Bullock merged with the Franklin Flexible Tractor Company in 1920.

CREEPING GRIP.

1915

**Campco Tractors, Inc.
Stockton, California**
Campco is another of those elusive companies for which little information can be found; in the research for this book, the trademark is the only extant data.

1925

**J. I. Case Company
Racine, Wisconsin**
(Earlier known as J. I. Case Threshing Machine Company)
J. I. Case Company goes back to 1842. This was an entirely different firm than the J. I. Case Plow Works; both were founded by J. I. Case, but as separate entities. The Case Eagle trademark predominated as the company's standard bearer for many years. The company is now known as Case-IH.

1909

1898

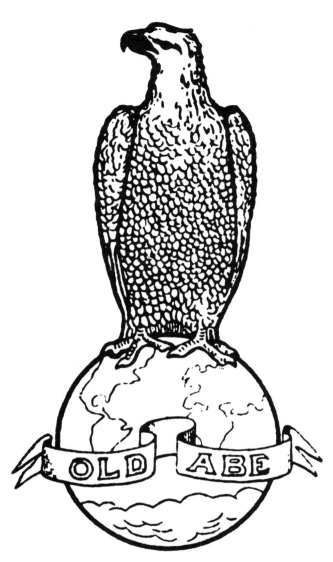

1915

1928

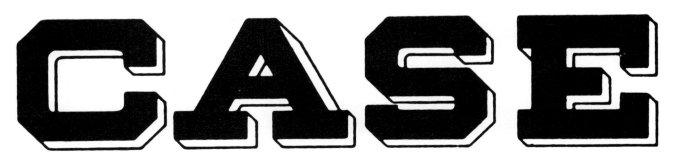

1930

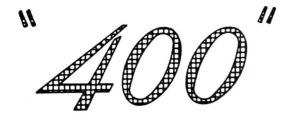

1955

1955

1957

CASE
1957

Case-o-matic
1958

J. I. Case Plow Works
The J. I. Case Plow Works and the Wallis Tractor Company operated from the same plant facilities, and in later years, the Wallis tractors were manufactured by the Plow Works. In 1928, this firm was bought out by Massey-Harris.

J.I. CASE
1915

CASE
1915

**Caterpillar Tractor Company
Peoria, Illinois**
After the Holt and Best merger of 1925 that formed Caterpillar, the company operated plants at San Leandro, California, and Peoria, Illinois. Today, Caterpillar has numerous manufacturing facilities, and is headquartered at Peoria.

LA CHENILLE

1925
French version of Caterpillar trade name

ГУСЕНИЦА

1927
Russian version of Caterpillar trade name

HOUSENKA

1927
Czechoslovakian version of Caterpillar trade name

RAUPE

1927
German version of Caterpillar trade name

CATERPILLAR

1936

КАТЕРПИЈЛАР

1927
Russian version of Caterpillar trade name

CAT

1949

**Central Tractor Company
Greenwich, Ohio**

The Centaur first appeared in 1921 as a motor-driven garden tractor. Numerous models were built during the next thirty years. The company also built small farm and industrial tractors.

CENTAUR

1921

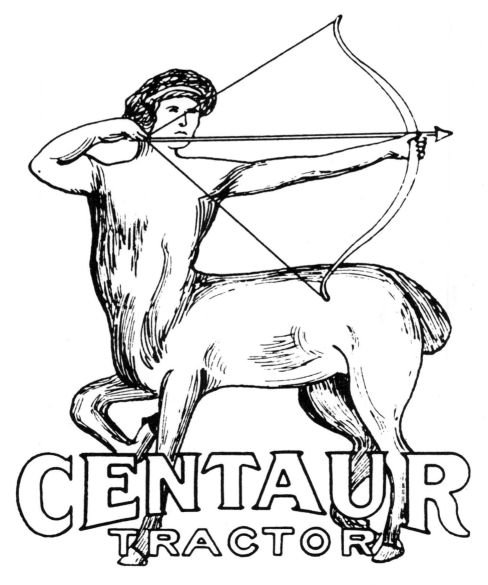

1923

Challenge Tractor Company
Minneapolis, Minnesota

Virtually nothing is known of the Challenge tractors except for its trademark, which claimed first use on October 1, 1916.

Andre Citroen
Paris, France

Although this firm is primarily known for its automotive line, the company also built tractors and farm equipment. This mark was first used in 1920, and it is possible that a small number came into the United States.

1916

1924

Clark Tructractor Company
Buchanan, Michigan

All indications are that the Clark Tructractor first appeared about 1919. Initially, these were an industrial tractor, as opposed to the ordinary farm tractor. From these beginnings, Clark went on to develop and manufacture a broad line of industrial equipment.

1923

33

TRUCTRACTOR

1923

DUAT

1925

CLARKAT

1927

TRUCTRACTOR

1927

Cleveland Tractor Company
Euclid, Ohio
Cleveland began building tractors in 1917 and developed an auspicious line of crawler tractors over the years. Its famous Cletrac crawlers were merged into Oliver Corporation in 1944.

Cleveland
1918

Cletrac   Cletrac
1919   1920

"Snap" Cletrac "Oiling System"
1926

Cletrac
CRAWLER
1929

The General
1939

Cockshutt Plow Company, Ltd.
Brantford, Ontario
Cockshutt Plow Company had beginnings as far back as 1877, but was absorbed by the White Farm Equipment line in 1962. Although the Cockshutt line was widely sold in Canada, it was also marketed in parts of the United States.

COCKSHUTT
1947

1949

Continental Company
Springfield, Ohio

Beginning in 1925, and perhaps earlier, the Cultor was developed as a small garden tractor. This firm was apparently one of many casualties in the Great Depression.

CULTOR

1926

Craig Tractor Company
Euclid, Ohio

A 15-25 tractor was marketed by Craig from 1918 to 1921. Due to the intense competition of the time, coupled with the postwar depression of the early 1920s, Craig made a well-intentioned but unsuccessful attempt to garner a portion of the tractor business.

1919

Daimler-Benz
Stuttgart, Germany
The Unimog was an attempt at a universal machine that could function as stationary power unit, tractor, truck, or for other duties when properly equipped.

UNIMOG
1957

Dearborn Motors Corporation
Highland Park, Michigan
While this trademark is specifically for grain drills and related equipment, the Dearborn Motors Corporation name was widely recognized in the tractor industry for several years.

DEARBORN PEORIA
1950

Virgil F. Deckert
Minneapolis, Minnesota
Little is known of the Pullet tractor, except for the manufacturer's claim that "she's a bird to pull." The Pullet trade name is absent from the extant tractor directories.

1918

Deere & Company
Moline, Illinois
With a history going back to 1837, Deere & Company was in the plow business for seven decades before getting into the tractor trade. Thus, several Deere trademarks appear that are not strictly related to farm tractors. Deere & Company continues its long tradition today as a major force in the farm equipment industry.

RED JACKET

1892
Plow trade name

1906

DEERE
1906

DEERE & MANSUR
1906

TRACTIVATOR.
1917

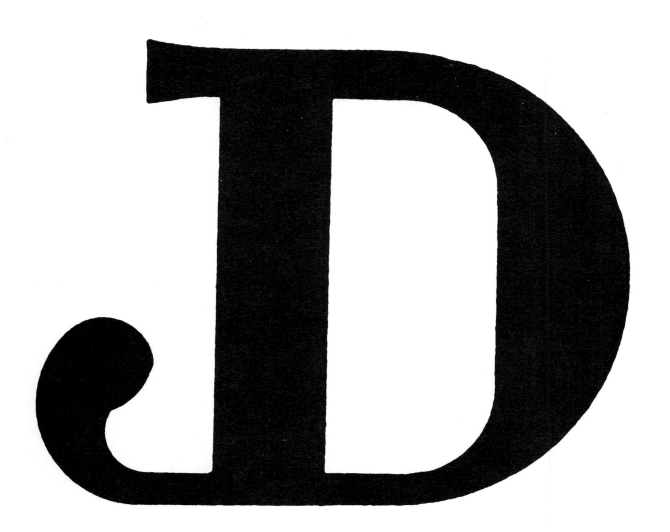

1920

40

1920

POWERFARMER

1927

1945

DEERE

1946

ROLL·O·MATIC

1947

JOHN DEERE
QUALITY FARM EQUIPMENT

1948

JOHN DEERE
hydraulic
POWR-TROL

1949

42

1945

**Detroit Harvestor Company
Detroit, Michigan**
Detroit Harvestor Company was an innovative firm that built many different machines ahead of the times. This company developed fully mounted tractor mowers, among other things. The Farmford was a conversion unit intended to convert a Ford Model T into a small tractor.

FARMFORD

1927

**Detroit Truck Company
Detroit, Michigan**
The Tonford was an attachment for converting the Ford Model T into a small truck, as well as serving various farm needs. Like the other conversions of its day, the life of this development was quite limited.

1916

Dixieland Motor Truck Company
Texarkana, Texas
Beginning about 1918, the Dixieland 12-25 tractor appeared. Although it was probably as good or better than its competitors, this tractor does not appear in the trade directories after 1920.

DIXIELAND

1919

Duplex Printing Press Company
Battle Creek, Michigan
In 1935, the trademark shown here was first applied to the Co-op tractors. In the next few years, a confusing maze of companies either sold or manufactured the Co-op designs. It appears that Duplex Machinery Company was one manufacturer, and perhaps as a successor, the Co-op was then made by the Cooperative Manufacturing Company.

CO-OP

1935

Clyde J. Eastman
Los Angeles, California
This trademark was filed in 1908, although first use was claimed for January 1902. Except for the trademark, nothing further has been located of this company.

FARMOBILE

1908

Ellinwood Industries
Los Angeles, California

Bear Cat garden tractors first emerged in 1946 from Ellinwood Industries. The company continued building garden tractors for several years.

1946

Emerson-Brantingham Company
Rockford, Illinois

Although organized in 1909, Emerson-Brantingham Company had roots going back to the 1850s. In the 1910-1914 period, Emerson-Brantingham acquired numerous equipment lines, including Geiser, Big 4, and others. Although a major manufacturer, financial problems forced the company to sell out to J. I. Case in 1928.

GEISER

1913

PEERLESS.
1913

NEW PEERLESS
1913

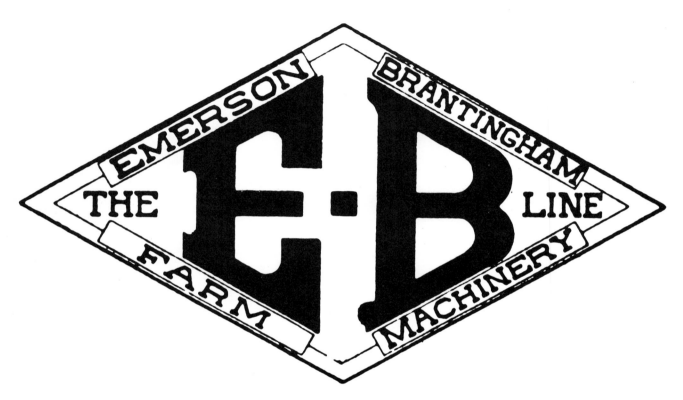

1916

LIBERTY
1917

Equipment Corporation of America
Wilmington, Delaware

No information has been located on this firm except for the trademark shown here.

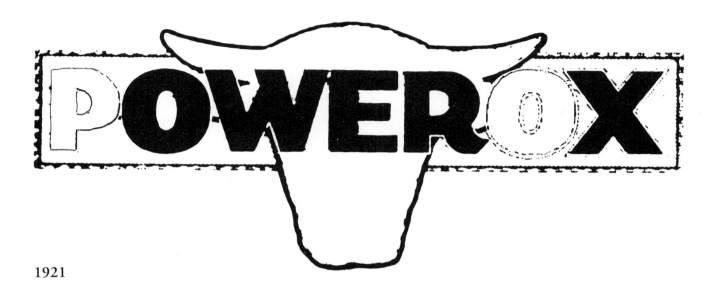

1921

A. J. Ersted
San Francisco, California

From the trademark drawing, the Auto Cat was obviously a small half-track crawler design. Except for this information, no data can be located on the 1930s design.

1933

**Euclid Road Machinery Company
Euclid, Ohio**
Euclid has built large industrial tractors, as compared to farm tractors. The Euclid name is virtually a household word in the construction industry.

1937

EUCLID

1948

Evans & Barnhill, Inc.
San Francisco, California
Research for this book has been unable to determine whether Evans & Barnhill functioned as a tractor manufacturer, or whether this firm might have been a large farm equipment distributor. Regardless, this mark was first used in 1923 on its equipment.

1923

Fairbanks, Morse & Company
Chicago, Illinois

Fairbanks, Morse began building internal-combustion engines in the 1890s, and first offered a tractor about 1910. During the next decade the company built a few different tractor models, but apparently preferred to devote its major research to the engine business. The company remains in the engine business with a factory located at Beloit, Wisconsin.

1912

FAIR-MOR

1918

1932

**Farm & Home Machinery Company
Orlando, Florida**

The Farmco tractors and implements, particularly the company's activities in the tractor business, remain somewhat of a mystery. The only evidence to date is this trademark, indicating first use in 1939.

1948

**A. B. Farquhar Company, Ltd.
York, Pennsylvania**

The Farquhar operations go back at least to 1870. In the following years, the company built steam traction engines, among many other items. About 1915, Farquhar entered the tractor business for a short time. Farquhar was sold to Oliver Corporation in 1952.

1925

51

FARQUHAR

1945

Ferguson-Sherman Manufacturing Corporation
Dearborn, Michigan

Harry Ferguson was located at Evansville, Indiana, at one time; an early trademark indicates that the name was applied to plows as early as 1911. Ferguson-Sherman, Inc., was at Evansville, Indiana, subsequently. The firm then moved to Dearborn, Michigan.

1923

1926

Ferguson

1940

Fiat s.A.
Turin, Italy
Fiat was a pioneer agricultural equipment builder. Although the early efforts were aimed at Italy and throughout Europe, the Fiat line eventually gained significance in the United States.

FIAT

1950

Food Machinery & Chemical Corporation
San Jose, California
The FMC tractor line grew largely from the acquisition of the Bolens garden tractor line. From the original Bolens Husky, FMC went on to build a large variety of tractors and implements.

1949

Versa-matic

1956

Ford Motor Company
Highland Park, Michigan
Henry Ford's interest in building a small farm tractor has been extensively documented, with his efforts finally resulting in the Fordson tractor. Some twenty years later came the famous Ford 9N, 2N, and 8N models, equipped with a three-point hitch system. From the later foundation, Ford has built an extensive line of farm tractors.

1915

Fordson

1917

1940

1943

1957

Four Wheel Traction Company
New York, New York
In the 1930s, the Super-4-Drive tractor appeared, albeit for only a short time. Little information can be located concerning this erstwhile entry into the tractor business.

1930

Fox River Tractor Company
Appleton, Wisconsin

While Fox forage harvesting equipment has had a wide reputation over the years, it is not generally known that Fox also built a farm tractor in 1919. Despite its neat appearance and rugged construction, the company opted to concentrate its efforts on its already established line of forage equipment.

1925

Franklin Tractor Company
Middletown, Ohio

Franklin developed its Franklin Flexible 15-30 tractor about 1920. Within a short time the firm merged with Bullock Tractor Company at Chicago to form the Franklin-Bullock Tractor Company.

1920

Frick Company
Waynesboro, Pennsylvania

Established in 1853, Frick Company adopted the "Eclipse" trademark as early as 1876. The company built an extensive line of steam engines in stationary and traction models, along with threshing machines, refrigerating equipment, and many other items. The Frick tractors were built in the decade following 1918.

1921

G.F.H. Corporation
Denver, Colorado

A large number of tractor manufacturers were in business for less than a year, and many more were in business for less than five years. One such short-lived venture was the Jerry tractor from G.F.H. Except for the trademark, virtually no information can be located.

1921

Gaar, Scott & Company
Richmond, Indiana

The famous Gaar-Scott Tiger trademark was filed in 1907. The company filed virtually the same mark for three different categories, as shown under this heading. Gaar-Scott was bought out by M. Rumely Company in 1913.

1907

Gamble-Skogmo Incorporated
Minneapolis, Minnesota

In the late 1940s, the Gamble's Stores opted into various farm equipment, including tractors. This excursion was a rather short one, but a few tractors were offered with the Farmcrest logo.

Gas Traction Company
Minneapolis, Minnesota

This company was one of the first to use a four-cylinder engine in a farm tractor. Beginning in 1910 the trademark shown here was put into use for the Big 4 tractor line. Emerson-Brantingham bought out the Big 4 in 1912.

1911

General Ordnance Company
Derby, Connecticut

The G-O tractor was built at Cedar Rapids, Iowa, in the factory previously owned by the Denning Tractor Company. Despite the financial backing of General Ordnance, the G-O tractor venture failed within a short time.

1919

General Implement Company of America, Inc.
Cleveland, Ohio

While this firm is known to have offered a sizeable line of farm equipment, it remains uncertain whether General Implement actually offered a tractor under its logo.

1943

General Tractor Company
Cleveland, Ohio

No information can be located on this firm, except for the trademark shown here; it indicates first use of the mark on October 1, 1927.

GENERAL

1928

General Tractor Company
Seattle, Washington

The Westrak tractor line emerged for but a short time from this factory, with the mark being first used on April 5, 1948. Little else is known of the company activities, nor have any illustrations appeared of this tractor.

WESTRAK

1948

Geneva Tractor Company
Geneva, Ohio

The Adapto-Tractor was an attachment for converting an automobile into a tractor. Numerous companies were engaged in this business for a few years, with the Ford Model T being a particular favorite for being converted thus.

ADAPTO-TRACTOR

1918

Gibson Manufacturing Corporation
Longmont, Colorado

Beginning in 1943, the Gibson tractor line gained considerable attention and limited popularity for a few years. The line faded during the 1950s.

1949

1949

Graham Bros.
Evansville, Indiana

Graham Bros. filed this trademark on May 13, 1920. For reasons unknown, the company's efforts in the tractor business do not appear to have been highly driven. In the late 1930s, the Graham-Bradley tractors appeared, but they were built by Graham-Paige Motors Corporation at Detroit, Michigan.

1920

Gray Tractor Company
Minneapolis, Minnesota

Gray Tractor Company used a unique drum-drive design on its tractors. Despite heavy competition and heavier odds, the company was fairly successful for a number of years. During the mid-1920s, however, Gray Tractor Company disappeared from the scene.

1923

1919

Hackney Manufacturing Company
St. Paul, Minnesota

The Hackney was of the autoplow design. In other words, this unique tractor carried the plows beneath the chassis. Luxurious seats as found in fine coaches were provided for the operator, and when this machine doubled as a road vehicle, additional seating was available. The Hackney was built for a short time, beginning about 1912 and ending about 1919.

1913

Hadfield-Penfield Steel Company
Bucyrus, Ohio

Hadfield-Penfield Steel Company offered various kinds of crawler tractors, beginning about 1918, and continuing at least into the 1920s. This company opted for road and industrial tractors, rather than for the ordinary farm tractor.

ONE MAN GRADER

1923

crawlerize

1926

RIGID RAIL

1925

ROAD HOG

1927

Paul Hainke Manufacturing Company
Hutchinson, Kansas

The Multivator appeared about 1950; it was apparently a small hand plow or engine-powered garden tractor. Virtually nothing further has been located outside of this trademark.

MULTIVATOR

1950

65

Happy Farmer Tractor Company
Minneapolis, Minnesota

Happy Farmer Tractor Company was organized in November 1915 at Minneapolis. By late the following year, the LaCrosse Tractor Company at LaCrosse, Wisconsin, was organized, with production of the Happy Farmer being moved to that city. Production at LaCrosse continued for several years.

Hart-Parr Company
Charles City, Iowa

Hart-Parr was the first company solely dedicated to building tractors, and as such, laid good claim to being "founders of the tractor industry." The company operated at Charles City until 1929 when it entered a merger forming the Oliver Farm Equipment Company.

HART-PARR

1920

Hebb Motor Company
Lincoln, Nebraska

No information can be found concerning the Patriot tractor from Hebb Motor Company, except for this trademark application filed in April 1918.

PATRIOT

1918

Heinz & Munschauer
Cleveland, Ohio

Leader tractors emerged toward the end of World War II, and remained on the market for several years. The postwar period saw many new and innovative designs, some of them borrowed from technologies developed during, or because of, that mighty conflict.

L
E
A
D
E
R

1948

Hession Tiller & Tractor Corporation
Buffalo, New York

The Wheat tractor came onto the market in 1919, but a similar tractor emerged two years earlier, and was simply sold under the Hession trade name. After 1920, the company was known as Wheat Tiller & Tractor Corporation.

Holt Manufacturing Company
Stockton, California

Benjamin Holt and the Holt Manufacturing Company had a history going back to about 1883. This company was early to develop steam power with large traction engines, and likewise pioneered the development of crawler tractors. Holt and Best merged in 1925 to form Caterpillar Tractor Company.

CATERPILLAR
1910

CATERPILLAR
1911

CATERPILLAR
1916

CATERPILLAR TIMES
1916
Monthly periodical trade name

69

CATERPILLAR

1917

CATERPILLAR

1919

ORUGA

1920
Holt tractor and internal-combustion engine trade name

HOLT

1920

Frank G. Hough, Inc.
Libertyville, Illinois

While not a farm tractor builder, the Hough Company was instrumental in developing a broad series of industrial tractors and related equipment, including tractor loaders and power shovels.

1953

1953

PAYLOGGER

1957

Hume-Love Company
Garfield, Washington
No information has been located regarding these tractor-powered machines, designed especially for the cultivating and harvesting industries. All apparently had their inception in 1938.

PoweRRoweR

1938

MotoRRoweR

1938

TRACTOR POWER

1938

Intercontinental Manufacturing Company
Dallas, Texas

Intercontinental entered the tractor business shortly after World War II, continuing into the mid-1950s. After that time the company apparently dropped out of the tractor business.

INTERCONTINENTAL

1949

1949

International Harvester Company
Chicago, Illinois

International Harvester itself dates back to 1902, it being the result of a consolidation of McCormick, Deering, Milwaukee, and several other companies. McCormick itself went back to the 1830s and the invention of the reaper by Cyrus Hall McCormick. By the early 1900s, the company was heavily into development of a gasoline tractor. By 1910, some proven designs were in the field, and in the 1920s, International Harvester announced the world's first commercially successful row-crop tractor, the Farmall. During the 1980s, IH merged with J. I. Case Company to form Case-IH.

1904
Trademark facsimile signature of James Deering

1904
Trademark facsimile signature of Cyrus H. McCormick

1904
Trademark facsimile signature of William H. Jones, vice president

McCORMICK

1913

DEERING

1913

MOGUL

1913

OSBORNE

1913

GIANT

1914

TITAN

1914

MOGUL

1914

TITAN

1917

MOGUL

1917

INTERNATIONAL

1917

The International Trail

1919
Monthly periodical trade name

The HARVESTER WORLD

1919
Monthly periodical trade name

DEERING

1920

McCORMICK

1920

McCORMICK-DEERING

1920

INTERNATIONAL

1920

1920

1920

1921 # McCORMICK

1921 # DEERING

1922 # FARMALL

1923 # McCORMICK-DEERING

1929 # FAIRWAY

1939 # LIFT-ALL

1944

1944

HI-SPEED FARMALL CUB
1947   1947

1947

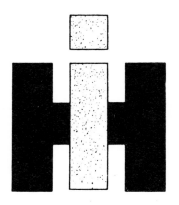

1954

FARMALL

1954

CUB

1956

**Interstate Engine & Tractor Company
Waterloo, Iowa**

The Plow Boy tractor made its first appearance in 1916. The company was organized by Sandy McManus who was involved with various other business ventures at the time. For inexplicable reasons, the Interstate Company disappeared shortly after it was organized.

Plow Boy

1917

Isaacson Iron Works
Seattle, Washington

Isaacson claimed first use of the Farm Dozer trademark in November 1939. However, no information on this company has been located in any of the tractor trade directories.

FARM DOZER

1940

J.T. Tractor Company
Cleveland, Ohio

During 1918, the J-T crawler tractor appeared on the market. This company was at least moderately successful, since various J-T models appear in the trade directories until at least 1928.

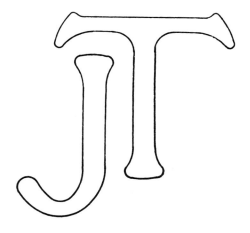

1920

Joliet Oil Tractor Company
Joliet, Illinois

The Steel Mule first appeared in 1914, and soon became one of the most popular of the small crawler tractors. In 1919, Joliet merged with the Bates Tractor Company, Lansing, Michigan, to form Bates Machine & Tractor Company. The latter firm built the Bates Steel Mule for several years.

STEEL MULE

1914

Horace Keane Aeroplanes, Inc.
New York, New York

Since the Ace tractor was ostensibly built by an airplane manufacturer, it seems entirely possible that the subject of this inquiry might have been an industrial or airport tractor. Except for this trademark application, nothing further has been ascertained about the tractor or its builder.

ACE

1920

Knickerbocker Motors, Inc.
New York, New York
The Knickerbocker was an attachment for converting an automobile into a tractor. In many instances, the plan was to convert to a tractor during the day, or for the week, and convert back to a car just in time to go to church or a family outing. Knickerbocker also built a small 5-10 tractor for a time.

KNICKERBOCKER

1917

John Minor Kroyer
Stockton, California
The Cliff tractor remains virtually unknown to tractor historians, with the 4-Pull model faring little better. The latter was built at San Pedro, California, from 1919 to 1924. At that time the company was acquired by Wizard Tractor Company of Los Angeles, California.

4 PULL

1917

CLIFF

1917

Landmaster, Ltd.
Hucknall, England
Landmaster apparently began using this trade name in 1950, and put it into commercial use during 1952. Although the United States sales for this British firm were relatively small, the 1950s marked the beginning of major tractor imports into the United States.

1957

**La Plant-Choate Manufacturing Company
Cedar Rapids, Iowa**

This company built a wide range of construction machinery, first with various kinds of scrapers and carriers, and later on with motor-propelled equipment. In the early 1950s, La Plant-Choate was taken over by Allis-Chalmers.

1931

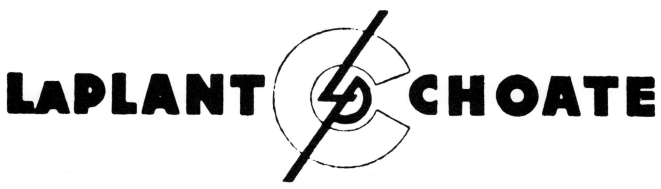

1948

Le Roi Company
West Allis, Wisconsin

Centaur garden tractors were just part of the overall line for this company. Also included was a virtual plethora of specialized implements for the truck gardener and orchardist. Le Roi was also a major engine builder.

1942

CENTAUR

LeTourneau-Westinghouse Company
Peoria, Illinois

R. G. LeTourneau Incorporated operated at various locations, including Stockton, California, and Longview, Texas. The company specialty was construction machinery; its tractors were built primarily for this purpose, rather than for agricultural uses.

1943

TOURNAPULL

1946

LETOURNEAU

1946

LETOURNEAU
LET
EQUIPMENT

TOURNATRACTOR
1954

LETOURNEAU
1957

PAK-RAPID
1959

SPEEDPULL
1959

TOURNAPULL
1960

Line Drive Tractor, Inc.
Chicago, Illinois
Current research fails to determine whether this was the same company listed at Milwaukee, Wisconsin. The Line Drive Tractor Company at that location was acquired by the Automotive Corporation of Fort Wayne, Indiana, in 1917.

1917

Lion Tractor Company
Minneapolis, Minnesota
The Lion tractor was intended to be a competitor with the Bull tractor, built in the same vicinity, and using the same general design. However, since the design would soon be relegated to obsolescence, both companies operated for only a few years. The Lion disappeared by 1918.

1915

1915

**McKinney Traction Cultivator Company
St. Louis, Missouri**
The McKinney was one of the first motor cultivators ever built, and dates back to experimental models in 1910, and full production a year later. This machine was ahead of its time, and the company was out of business within a few years.

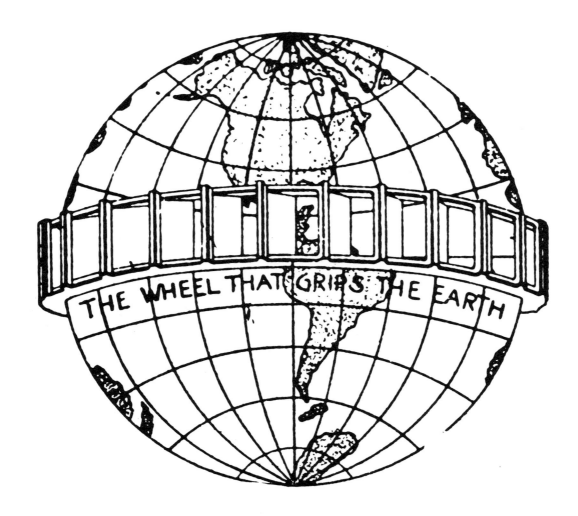

1912

87

M.B.M. Manufacturing Company
Milwaukee, Wisconsin
This company built the famous Red-E line of garden tractors and power lawn mowers. Innovative, rugged, and reliable, the Red-E line carved a place in tractor history with numerous examples still surviving among vintage machinery collectors.

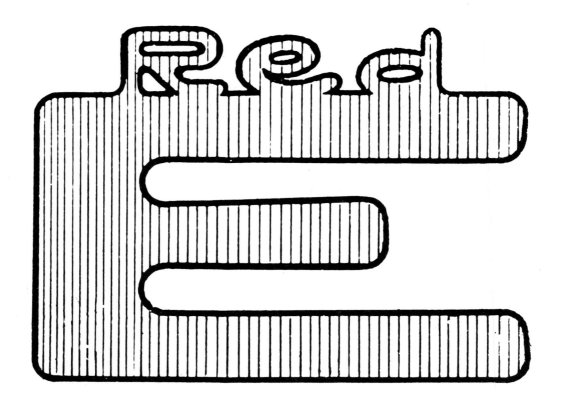

1924

Marsh-Capron Manufacturing Company
Chicago, Illinois
The Shop Mule was an early example of an industrial tractor. Although tractors are commonly thought of as a piece of farm equipment, there was also a tremendous need for shop tractors in manufacturing establishments; this tractor was designed specifically for that purpose.

1918

Marvel Tractor Company
Columbus, Ohio

Marvel tractors, despite an attractive trademark, do not appear to have gone beyond a few prototype copies. Perhaps it was the withering competition of the time, or perhaps it was from other causes, but the Marvel is a typical example of a venture that really never got out of the gate, much less placed or showed.

1921

Maschinenfabrik Augsburg-Nurnburg A.G.
Nurnberg, Germany

Fuel-Master tractors were primarily of diesel design, especially since M.A.N. was a world leader in diesel engine development. A limited number made their way to the United States, but the line was quite popular in Europe.

FUEL-MASTER

1957

Massey-Harris Company
Racine, Wisconsin

Massey-Harris resulted from an 1891 merger, but this Canadian company never got into the tractor business in a big way until 1928. With its acquisition of the J. I. Case Plow Works, and its Wallis tractors, Massey-Harris went on to develop an extensive tractor and farm equipment line. The company later merged with Ferguson to form the Massey-Ferguson line.

1936

CHALLENGER

1936

Pony

1949

1951

90

Colt

1952

Pacer

1953

WORK BULL

1956

1958

Massillon Engine & Thresher Company
Massillon, Ohio

The trademark shown here was first used in 1894, but this company may have been in operation prior to that time. Unfortunately, no further information can be located for this company.

1894

Mayer Brothers Company
Mankato, Minnesota

Mayer Brothers began building tractors in 1914, and in 1918, the company name was changed to Little Giant Company. Its high-quality tractors remained on the market until about 1927, and even enjoyed a considerable export trade during the production period.

1916

1922

Midland Company
South Milwaukee, Wisconsin
The Dandy Boy was but one of many different tractor makes appearing after World War II. Tractor production practically ceased for the duration, leaving the market wide open for a few years.

DANDYBOY

1949

**Midwest Engine Company
Indianapolis, Indiana**
Midwest Engine began building small tractors about 1918, with its Utilitor tractor gaining immediate and long-lasting fame. This small garden tractor remains a favorite among collectors of vintage garden tractors.

1919

1919

1919

1919

Miller Traction Tread Company
Chicago, Illinois
The TracTred design was essentially a conversion unit designed to make a "half-track" out of an ordinary farm tractor, especially the Fordson. Fairly popular for a time, conversion kits like this one eventually faded away, in favor of a dedicated design.

TracTred

1922

Minneapolis-Moline Power Implement Company
Minneapolis, Minnesota
Minneapolis-Moline resulted from a 1929 merger. This grouping of trademarks also includes some originally issued to the Moline Plow Company of Moline, Illinois. The latter was a major player in the farm tractor business with its Moline Universal tractor, a unique row-crop design of the 1918-1923 period.

UNIVERSAL

1917
Moline Plow Company traction engine trade name

MOLINE

1920

UNI-TILLER

1923

1904
Moline Plow Company "Dutchman"

1931

COMFORTRACTOR

1945

MOLINE

1946

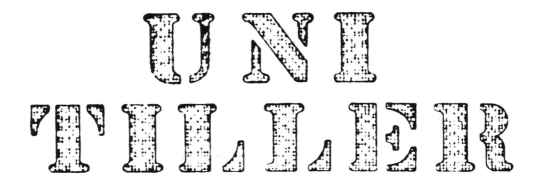

1947

MINNEAPOLIS-MOLINE

1950

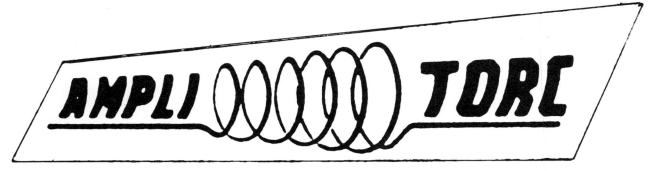

1956

1960

Minneapolis Steel & Machinery Company
Minneapolis, Minnesota

Minneapolis Steel was the manufacturer of the famous Twin City tractor line. This broad-based company was also a manufacturer of heavy stationary steam engines and gas engines. In 1929, it merged with Moline Plow Company and others to form the Minneapolis-Moline Power Implement Company.

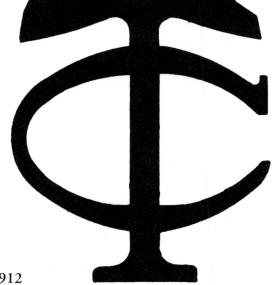

1912

1914

1920

**Mustang Motorcycle Corporation
Glendale, California**
No information whatsoever has surfaced regarding the Mustang tractor from this company. Presumably it was a small tractor, perhaps even a garden or lawn tractor design.

1953

**National Cooperatives, Inc.
Chicago, Illinois**
National Cooperatives emerged in the 1930s as a supplier of most kinds of farm equipment and farm supplies. The concept was to use cooperative buying to eliminate the "middleman" and pass the savings onto the members. Co-op tractors appeared at various times.

1905

**National Implement Company
Princeton, New Jersey**
The Power Horse tractors first appeared about 1948 and were marketed for a short time thereafter. This tractor featured a unique four-wheel-drive design.

POWER HORSE

1950

NATIONAL POWER HORSE

1953

National Tractor Company
Chicago, Illinois

National tractors were built in Cedar Rapids, Iowa. The company was a reorganized successor to General Ordnance Company, which earlier built the G-O tractor. The latter in turn, was the successor to the Denning Motor Implement Company of Cedar Rapids. National tractors were built as late as 1919.

1918

New Britain Machine Company
New Britain, Connecticut

During 1920, New Britain announced a new tractor-cultivator. Offered in two different sizes, it remained on the market for only several years.

1920

New Idea Spreader Company
Coldwater, Ohio

New Idea began building manure spreaders as early as 1899. The company remained as a short-line builder, specializing in implements, rather than going into the tractor business. During the 1950s, New Idea acquired the "Uni-System" line developed by Minneapolis-Moline.

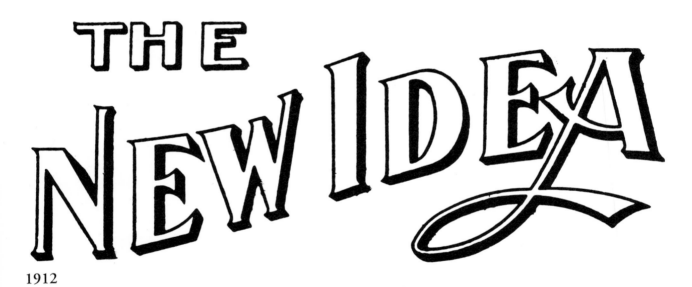

1912

New Way Motor Company
Lansing, Michigan

Although New Way has long been identified as a gasoline engine builder, the company apparently had plans of going into the tractor business, and may have even built a limited number, probably using its own unique air-cooled engines as the prime mover. New Way continued in the engine business for many years.

1911

Nilson Tractor Company
Minneapolis, Minnesota

Nilson was organized in 1913 and operated until 1929. Its tractor designs were always unique, and given the longevity of the product line, must have been satisfactory to numerous farmers. At least four different models were built.

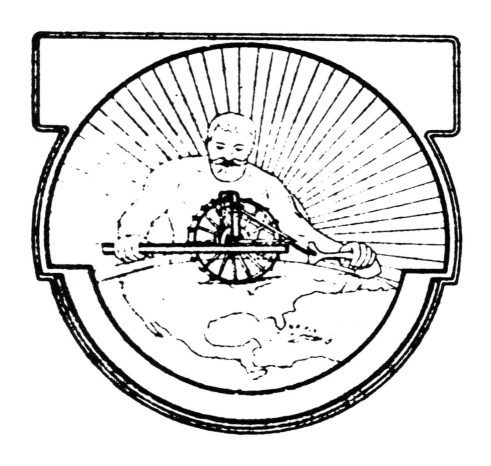

1918

Oliver Corporation
Chicago, Illinois

Oliver Farm Equipment Company was formed in 1929 with the merger of Oliver Chilled Plow Company, Nichols & Shepard, Hart-Parr Company, and the American Seeding Machine Company. Eventually the name was changed to Oliver Corporation. In 1960, the firm was merged into White Farm Equipment.

OLIVER

1920

1929

1929

Hydra-lectric
1950

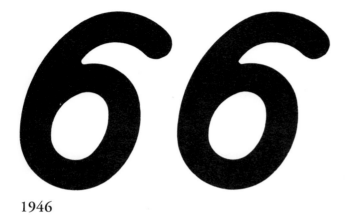
1946

1946

Oliver Tractor Company
Knoxville, Tennessee

Beginning in 1917, the Oliver crawler tractor appeared, remaining on the market until about 1920. Rated at 15 drawbar horsepower, it was a small tractor with an attractive appearance. This company was not related to Oliver Farm Equipment Company in any way.

1917

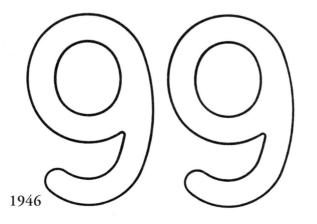
1946

One Wheel Truck Company
St. Louis, Missouri

The Autohorse was apparently a style of industrial tractor designed for moving cargo from place to place. However, aside from this trademark, no further information has been located.

Autohorse

1917

Pan Motor Company
St. Cloud, Minnesota

Pan Motor Company became an entity during 1916, and over the next two years endeavored to manufacture its products, the Pan Tank-Tread tractor and the Pan automobiles. Neither venture succeeded, and Pan was out of business by 1920.

1919

James D. Park
Chicago, Illinois
No information outside of this trademark has been found for the Park tractor.

1919

Pioneer Tractor Manufacturing Company
Winona, Minnesota
Pioneer Tractor Company was incorporated on April 28, 1909. The company built several sizes of tractors in the following years, including a few copies of a gigantic 45-90 model. After reorganizing as Pioneer Tractors, Inc., in 1925, the company fell into oblivion shortly thereafter.

PIONEER

1911

The Pond Company
South Bend, Indiana
Harold E. Pond began the Pond Garden Tractor Company of Ravenna, Ohio. The "Speedex" trademark was first used in 1935. At an undetermined point, The Pond Company relocated to South Bend, Indiana, and continued with its Wheel-Horse garden tractors and gardening equipment.

SPEEDEX

1945

Wheel-Horse

1953

Post Tractor Company
Cleveland, Ohio

Incorporated in 1918, the Post Tractor Company only stayed in the market for a couple of years. The unique design was perhaps not to the liking of farmers; instead, they preferred the Fordson, the IHC Titan, and similar configurations.

1920

Rein Drive Tractors, Ltd.
Toronto, Ontario

Several companies built "rein-drive" or "line-drive" tractors. The rationale was that this design would be better accepted by farmers who had never before driven anything but horses or mules. The concept never gained great acceptance, despite the efforts of many inventors.

1918

Rototiller, Inc.
Wilmington, Delaware

The once-over concept of field cultivation gained favor with numerous rotary tiller designs. However, it was best suited to certain soil types and for specific crops. Thus the "Rototiller" had little success as an all-around farm tillage tool.

1932

M. Rumely Company
LaPorte, Indiana

This firm had a history going back to the 1850s when Meinrad Rumely began building grain threshers. The company flourished over the years, and probably reached its epitome with the 1909 introduction of the OilPull tractors. In 1915, the firm was reorganized as the Advance-Rumely Thresher Company, and in 1931, it was acquired by Allis-Chalmers.

1909

1909

1910

RUMELY GAS PULL
LA PORTE

1912

1912

1914

Russell & Company
Massillon, Ohio

Russell & Company began building grain threshers as early as 1842. Its trademark of "The Boss" was first used in 1881, and continued until the company was sold at auction in 1927. Russell built numerous tractor models beginning in 1909, and continued until the company ceased business operations.

1913

Samson Tractor Company
Janesville, Wisconsin
From its trademark application, it appears that Samson began building tractors as early as 1904. By 1914, the company had developed its unique Sieve-Grip design. About 1919, General Motors bought out the Samson line and moved it to Janesville, Wisconsin, where the Samson tractors were built for a few years.

1915

SAMSON

1920

Sears, Roebuck & Company
Chicago, Illinois
The significant history of the David Bradley trademarks is that they were acquired by Sears, Roebuck in 1910. Subsequently the company marketed the David Bradley line extensively, and at various times offered tractor models under the Bradley and other trademarks.

DAVID BRADLEY

1948

db DAVID BRADLEY

1958

Shaw Manufacturing Company
Galesburg, Kansas
During the 1920s, Stanley W. Shaw developed the famous Du-All tractors, a small design intended primarily for gardeners and orchardists. Shaw Du-All tractors remained on the market for many years.

DU-ALL

1949

Shaw-Enochs Tractor Company
Stillwater, Minnesota

By 1920, Shaw-Enochs had developed a small tractor built over a road grader. This was an early attempt at a motorized grader, and the company went on to build a variety of road-working machinery.

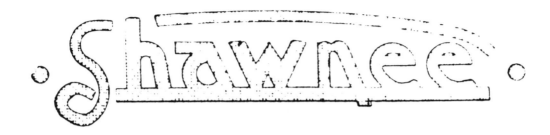

1920

THE CENTURY

1929

Simplicity Manufacturing Company
Port Washington, Wisconsin

Simplicity Manufacturing Company was organized in 1921 from the remains of the Turner Manufacturing Company. The latter had built farm tractors, and in 1937 Simplicity began building garden tractors for Montgomery Ward. Simplicity began building riding lawn tractors for Allis-Chalmers in 1961, and was bought out by the latter in 1965.

1952

WONDER–BOY

1956

"Quick-Hitch"

1957

Geo. H. Smith Steel Casting Company
Milwaukee, Wisconsin

This company developed the famous Trackson crawler attachment designed for the Fordson tractor. Its success led to other developments in the field, including various kinds of road-building machinery.

1924

1924

1932

Southern Motor Manufacturing Association, Ltd.
Houston, Texas

Southern began using the Ranger trade name on automobiles, motor trucks, and tractors in 1918, presumably about the same time the company began business. Its Ranger motor cultivator enjoyed limited success for a short time, and the Ranger automotive line remained until the company went into receivership during 1922.

1919

E. G. Staude Manufacturing Company
St. Paul, Minnesota

Staude designed and built a conversion unit to make a tractor out of an automobile. Occasionally these were seen with "Mak-A-Ford," "Mak-A-Tractor," and so on. Given the popularity of these conversion units for a few years, Staude was one of the leaders in this interesting field.

1917

Stroud Motor Manufacturing Association, Ltd.
San Antonio, Texas
In 1920, Stroud built an interesting tractor bearing an uncanny resemblance to the Farmall row-crop design. As such, the Stroud, if it had remained, might well have laid claim to being the first successful row-crop (tricycle) model. Unfortunately, the company was out of business before achieving that goal.

ALL-IN-ONE
1920

Toro Manufacturing Corporation
Minneapolis, Minnesota
Toro began building motor cultivators as early as 1919. The company also developed a line of lawn mowing tractors at this time, and continued by developing self-contained lawn mowers by 1925.

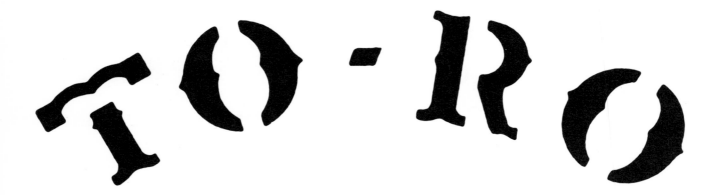

1920

1948

The Towmotor Company
Cleveland, Ohio
By 1919, Towmotor had designed an industrial tractor, and from this beginning went on to develop an extensive line of lift trucks and materials handling equipment.

TOWMOTOR
1921

1941

TOWMOTOR
1956

Tractomotive Corporation
Deerfield, Illinois
This company made its way in the tractor business by developing a wide variety of attachments for tractors, especially crawler tractors. The trade names shown here are indicative of the broad range of Tractomotive developments.

TRACTO-A SURE SIGN OF MODERN DESIGN
1957

TRACTOPARTS
1957

TRACTOSIDEBOOM
1957

TRACTOHOE
1957

Tractor Producing Corporation
New York, New York
No information has been located for this company outside of the trademark application shown here.

LIBERTY

1918

**Turner Manufacturing Company
Port Washington, Wisconsin**
As noted above under the Simplicity Manufacturing Company heading, Turner was the immediate predecessor. Turner built the Simplicity tractor for only a short time before being taken over in the Simplicity reorganization.

TURNER Simplicity

1918

**U. S. Tractor Corporation
Wilmington, Delaware**
Shortly after World War II, the USTRAC crawler tractors appeared for a few years, but little information has been located concerning the company's activities.

1949

United Tractor & Equipment Corporation
Chicago, Illinois

This firm was organized in 1929 by some 32 independent implement and machinery builders. Allis-Chalmers built the tractor, and for all practical purposes, it was the same as its Model U, even though it carried the United logo on the radiator.

1929

Unitractor Company
Indianapolis, Indiana

No information can be found on the Unitractor from this firm. Apparently it operated in the late 1930s and at least into the late 1940s.

1945

Vaughn Motor Works, Inc.
Portland, Oregon

Vaughn was well known for its log saws and other equipment, but virtually nothing is known of its Flex-Tred tractors which seem to have come onto the market about 1926.

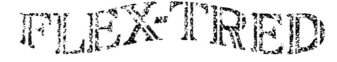

1927

Velie Motors Corporation
Moline, Illinois

Velie was an early builder of farm tractors, and in 1908 the Velie Motor Vehicle Company was formed to built automobiles. In the 1916-1920 period, Velie was particularly active in the farm tractor business.

1919

Wellman-Seaver-Morgan Company
Cleveland, Ohio

Wellman-Seaver-Morgan built the Akron tractor from 1918 to 1922. An unusual feature was the electric starting system, virtually unheard of for any tractor of this period.

1919

Charles S. Whitworth
Cedar Falls, Iowa

Early tractors were very difficult to steer. When used for plowing especially, steering was a tiring task. To this end then, the Dreadnought Guide was designed to ease the labor of steering.

Dreadnought Guide

1912

Wheel Horse Products, Inc.
South Bend, Indiana

Wheel Horse, according to this trademark application, offered a wide variety of implements, in addition to its own small tractors. Of course, the majority of these implements were designed especially for the company's own tractors.

1958

Wichita Falls Motor Company
Wichita Falls, Texas

The Wichita tractors were marketed in the 1911-1920 period. By the latter date, competition in the tractor industry was so intense that the majority of small tractor builders withered. Some were able to survive, some went on to other endeavors, and many were swept aside by the rushing tide.

1918

1920

Will-Burt Company
Orrville, Ohio

Little Farmer power cultivators first appeared after World War II, and remained on the market for a few years. Little else is known of this manufacturing venture.

1949

Winter Manufacturing Company
Tacoma, Washington
The Mighty Man garden tractors appeared concurrently with the end of World War II. No further information is available for this company.

MIGHTY MAN

1948

Wood Bros., Inc.
Des Moines, Iowa
The Wood Bros. Thresher Company built steam traction engines for a number of years, beginning about 1900. About 1946, the company was sold to Dearborn Motors Company, and in 1955 the factory became an implement plant for Ford Motor Company.

WOOD BROS.

1949

Index

Advance-Rumely Company, 6-7
Advance Thresher Company, 8
Albaugh-Dover Company, 8-9
S. L. Allen & Company, Inc., 9
Allis-Chalmers Manufacturing Company, 9-10
American Engine & Tractor Company, 10
American Tractor & Foundry Company, 11
American Tractor Corporation, 11
American Tractor Corporation, 12
Aro Tractor Company, 12
Aultman-Taylor Company, 12
Auto Power Company, 12
Autopower Company, 12
Avery Company, 13-14

B. F. Avery & Sons Company, 15-16
Badley Tractor Company, 16
Baines Engineering Company, 16
William Stimson Barnes, 17-18
Bean Spray Pump Company, 18
Belle City Manufacturing Company, 19-20
Beltrail Tractor Company, 20
C. L. Best Tractor Company, 20-22
Blumberg Motor Manufacturing Company, 22
David Brown Corporation, Ltd., 23
Buckeye Manufacturing Company, 23
Buffalo Pitts Company, 23-24
Bull Tractor Company, 24
Bullock Tractor Company, 24

Campco Tractors, Inc., 25
J. I. Case Company, 25-30
J. I. Case Plow Works, 30
Caterpillar Tractor Company, 4, 31-32
Central Tractor Company, 32
Challenge Tractor Company, 33
Andre Citroen, 33
Clark Tructractor Company, 33-34

Cleveland Tractor Company, 35-36
Cockshutt Plow Company, Ltd., 36-37
Continental Company, 37
Craig Tractor Company, 37

Daimler-Benz, 38
Dearborn Motors Corporation, 38
Virgil F. Deckert, 38
Deere & Company, 39-43
Detroit Harvestor Company, 43
Detroit Truck Company, 43
Dixieland Motor Truck Company, 44
Duplex Printing Press Company, 44

Clyde J. Eastman, 44
Ellinwood Industries, 45
Emerson-Brantingham Company, 45-46
Equipment Corporation of America, 47
A. J. Ersted, 47
Euclid Road Machinery Company, 48-49
Evans & Barnhill, Inc., 49

Fairbanks, Morse & Company, 50
Farm & Home Machinery Company, 51
A. B. Farquhar Company, Ltd., 51-52
Ferguson-Sherman Manufacturing Corporation, 52-53
Fiat s.A., 53
Food Machinery & Chemical Corporation, 53-54
Ford Motor Company, 54-55
Four Wheel Traction Company, 55
Fox River Tractor Company, 56
Franklin Tractor Company, 56
Frick Company, 57

G.F.H. Corporation, 57
Gaar, Scott & Company, 58
Gamble-Skogmo Incorporated, 59
Gas Traction Company, 59

General Ordnance Company, 60
General Implement Company of America, Inc., 60
General Tractor Company, 61
General Tractor Company, 61
Geneva Tractor Company, 61
Gibson Manufacturing Corporation, 61-62
Graham Bros., 62
Gray Tractor Company, 63

Hackney Manufacturing Company, 64
Hadfield-Penfield Steel Company, 65
Paul Hainke Manufacturing Company, 65
Happy Farmer Tractor Company, 66
Hart-Parr Company, 67
Hebb Motor Company, 67
Heinz & Munschauer, 67
Hession Tiller & Tractor Corporation, 68
Holt Manufacturing Company, 68-70
Frank G. Hough, Inc., 70-71
Hume-Love Company, 71-72

Intercontinental Manufacturing Company, 72
International Harvester Company, 73-80
Interstate Engine & Tractor Company, 80
Isaacson Iron Works, 81

J.T. Tractor Company, 81
Joliet Oil Tractor Company, 81

Horace Keane Aeroplanes, Inc., 81-82
Knickerbocker Motors, Inc., 82
John Minor Kroyer, 82

Landmaster, Ltd., 82-83
La Plant-Choate Manufacturing Company, 83
Le Roi Company, 84
LeTourneau-Westinghouse Company, 84-85
Line Drive Tractor, Inc., 85
Lion Tractor Company, 86-87

McKinney Traction Cultivator Company, 87
M.B.M. Manufacturing Company, 88
Marsh-Capron Manufacturing Company, 88
Marvel Tractor Company, 89
Maschinenfabrik Augsburg-Nurnburg A.G., 89-90
Massey-Harris Company, 4, 90-91
Massillon Engine & Thresher Company, 92
Mayer Brothers Company, 92-93
Midland Company, 93
Midwest Engine Company, 94
Miller Traction Tread Company, 95
Minneapolis-Moline Power Implement Company, 95-98
Minneapolis Steel & Machinery Company, 98-99
Mustang Motorcycle Corporation, 99

National Cooperatives, Inc., 99-100
National Implement Company, 100-101
National Tractor Company, 101
New Britain Machine Company, 102
New Idea Spreader Company, 102
New Way Motor Company, 102
Nilson Tractor Company, 103

Oliver Corporation, 103-105
Oliver Tractor Company, 105
One Wheel Truck Company, 105-106

Pan Motor Company, 106
James D. Park, 107
Pioneer Tractor Manufacturing Company, 107
The Pond Company, 107-108
Post Tractor Company, 108

Rein Drive Tractors, Ltd., 108
Rototiller, Inc., 109
M. Rumely Company, 109-112
Russell & Company, 112

Samson Tractor Company, 113-114
Sears, Roebuck & Company, 114
Shaw Manufacturing Company, 114
Shaw-Enochs Tractor Company, 115
Simplicity Manufacturing Company, 115
Geo. H. Smith Steel Casting Company, 116
Southern Motor Manufacturing Association, Ltd., 116-117
E. G. Staude Manufacturing Company, 117
Stroud Motor Manufacturing Association, Ltd., 118

Toro Manufacturing Corporation, 118
The Towmotor Company, 119
Tractomotive Corporation, 119
Tractor Producing Corporation, 119-120
Turner Manufacturing Company, 120

U. S. Tractor Corporation, 120
United Tractor & Equipment Corporation, 121
Unitractor Company, 121

Vaughn Motor Works, Inc., 121
Velie Motors Corporation, 122

Wellman-Seaver-Morgan Company, 122
Charles S. Whitworth, 123
Wheel Horse Products, Inc., 123
Wichita Falls Motor Company, 124
Will-Burt Company, 125
Winter Manufacturing Company, 126
Wood Bros., Inc., 126